Forces, Fields and Gravitational Field

Preface

Forces and fields are a great deal in Physics. In this book, we present the ideas of forces and fields in a simple-to-understand and interactive manner. Our intention is that students grasp the concept of forces in Physics and that they understand why fields are used to express forces in nature. Only gravitational forces/fields are extensively treated in this book, other forces/fields are treated extensively in other books in this series. For a more beautiful understanding, we have used a good number of relevant numerical examples/exercises to present the ideas.

FORCES, FIELDS AND GRAVITATIONAL FIELD

1 Forces: Definition

| 1 |

Force is an agent that causes motion. It can change the state of a body in uniform motion or at rest. Force is defined as the time rate of change of momentum of a body.

Types of forces: Contact forces and force fields.

Contact forces are forces that exist between surfaces in contact, example, tension, reaction force, and forces of push or pull, frictional forces.

Force fields are forces that act between two bodies at a distance. It does not require the two bodies to be in contact, unlike in contact forces. The gravitation force, electromagnetic, electric and magnetic forces are examples of force fields.

2 Fields.

| 2 |

Field in physics represents a region under the influence of some physical agency such as gravitation, magnetism and electricity. The concept of field is a convenient and informative method of describing the influence of one body over another body separated from the first by some distance. A Field is also a region between two bodies where action of force is experienced, the shorter the distance between the two bodies, the greater the strength of the field and vice versa. Here, the force is the force fields which do not require the two bodies to be in contact and it include examples in plan 1.

2.1 Classes of fields.

| 3 |

There are two classes of fields namely:

(i) **Scalar fields**: These are fields that have only magnitude but no direction. Examples of scalar fields are the distribution of temperature in space, density, pressure in fluid and energy.

(ii) **Vector fields**: These are fields that have both magnitude and direction. Examples of vector fields are gravitational, electric, magnetic and electromagnetic fields.

2.2 How to map the Fields

4

For scalar fields, their magnitudes are mapped by lines; the lengths of the lines indicate their magnitudes. For vector fields, both magnitudes and directions are specified or mapped by lines with arrow-heads on them; the lengths of the lines indicate their magnitudes while the arrow-heads indicate their directions.

A line of force is an imaginary line drawn in a field (electric, magnetic or gravitational) such that the direction (North pole to South pole or Positive to Negative) of the force gives the direction of the field.

3 The concept of gravitational field.

5

Gravitational field is a region or space around a mass in which the gravitational force of the mass can be felt. A gravitational field surrounds everybody that has mass, and this field fills up all space. When a body or an object is thrown up, it will come back to the ground because of the gravitational pull on the body as shown below.

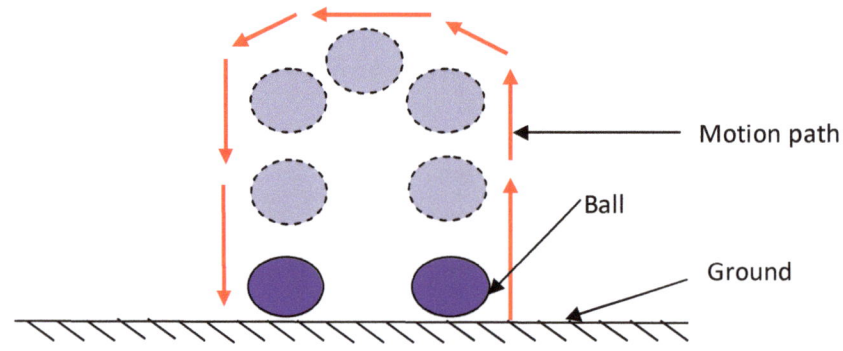

Figure 1. A body under the influence of Earth's gravity

3.1 Newton's law of universal gravitation.

The law of universal gravitation was put forward by Isaac Newton and it claims that gravity does not only act on the earth's surface but extends throughout the universe. Gravity or gravitational force is, therefore, the force of attraction between objects at some distance.

Newton's law of universal gravitation states that every particle in the universe attracts every other particle with a force that is proportional to the product of their masses and inversely proportional to the square of the distance between them. The force of attraction acts along the line joining the centers of the two bodies.

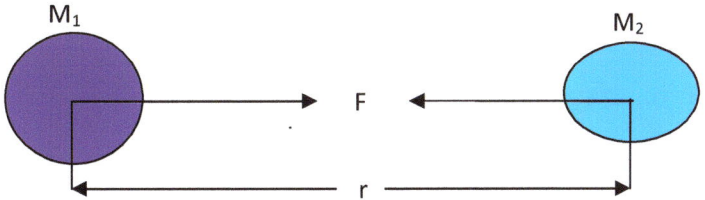

Figure 2. Force of attraction between two particles.

Can we move on? Yes!

3.2 Mathematical expression of the universal law

Following the definition above.

$$F \; \alpha \; \frac{M_1 M_2}{r^2} \; -------------------- \; 1$$

$$F = \frac{Gm_1 m_2}{r^2} \; -------------------- \; 2$$

Where m_1 and m_2 are the masses of the two particles, r is the distance between them and G is a universal constant of gravitation or simply gravitational constant.

$$G = 6.67 \times 10^{-11} kg^{-1} m^3 s^{-2} \; (G = 6.67 \times 10^{-11} Nm^2 kg^{-2})$$

If we have identical masses, $m_1 = m_2 = M$

$$F = \frac{GM^2}{r^2} \; ------------------- \; 3$$

3.3 Explanation:

The gravitational force of attraction form an action - reaction pair as in Newton's Third law. As shown in fig. 2, the mass m_1 is attracted towards the mass m_2 with a force F towards the right. At the same time, the mass m_2 is attracted by the mass m_1 with an equal force but in the opposite direction.

Now, we need not wonder or ask about what keeps the sun, moon and stars in space: what makes the moon revolve round the earth and the planets round the sun? The obvious answer is gravitational force.

Note: In this Law, $F = \dfrac{Gm_1m_2}{r^2}$, as distance between the particles increases, the gravitational force decrease and vice versa. Gravitational force is always attractive unlike electric force which can be attractive or repulsive.

Let us solve example, are you following?

Question 1

Calculate the force of attraction between two small objects of mass 10kg and 50kg respectively which are 10cm apart. Take G as $6.67 \times 10^{-11} Nm^2kg^{-2}$.

Hint: Convert 10 centimeters to meters and then solve; what did you get?

Convince yourself that $F = 3.335 \times 10^{-6} N$. And if you can't, move on to next plan.

Solution:

$G = 6.67 \times 10^{-11} Nm^2kg^{-2}$, $r = 10cm$, $F = ?$

If 100cm = 1m, then, 10cm =? $\Longrightarrow$ $\dfrac{10cm}{100cm} \times 1m = 0.1m$, $r = 0.1m$

m_1 and $m_2 = 10kg$ and $50kg$

force of attraction, $F = \dfrac{Gm_1m_2}{r^2}$

$$F = \frac{6.67 \times 10^{-11} \times 10 \times 50}{(0.1)^2} = 3.335 \times 10^{-6} \, N \, ans$$

Question 2

| 11 |

The earth has a mass of 6.0×10^{24} kg, the moon a mass of 7.4×10^{22} kg. Calculate the force of attraction between the two if the distance between their centers is 3.84×10^8 m.

Solution

| 12 |

$M_E = 6.0 \times 10^{24}$ kg, $M_m = 7.4 \times 10^{22}$ kg

$r = 3.84 \times 10^8$ m, $\quad F_{EM} = ?$

$$F_{EM} = \frac{GM_E M_M}{(r_{EM})^2} = \frac{6.67 \times 10^{-11} \times 6 10^{24} \times 7.4 \times 10^{22}}{(3.84 \times 10^8)^2}$$

$F_{EM} = 2.008 \times 10^{20} \, N \cong 2 \times 10^{20} N$

Question 3

| 13 |

The force of attraction between two point masses is 10^{-4}N when the distance between them is 0.18m. If the distance is reduced to 0.06m, calculate the force.
(A) 1.1×10^{-5}N (B) 3.3×10^{-5}N (C) 3×10^{-4}N (D) 9×10^{-4}N

[JAMB]

Solution !

| 14 |

Solution : $F = 10^{-4}$N, $r = 0.18$m

Substituting in the equation: $F = \frac{Gm_1 m_2}{r^2}$, we have

$$10^{-4} = \frac{Gm_1 m_2}{(0.18)^2}$$

$Gm_1m_2 = 10^{-4} (0.18)^2$

$Gm_1m_2 = 3.24 \times 10^{-6} \, Nm^2$.

When the distance between the two masses is reduced to 0.06. we have

$$F = \frac{Gm_1m_2}{(0.06)^2} = \frac{3.24 \times 10^{-6}}{(0.06)^2} = 9 \times 10^{-4} \, N$$

Question 4

15

The ratio of electro-static force F_E to gravitational force F_G between two protons each of charge (e) and mass (m) at distance (d) is

(A) $\dfrac{e^2}{Gm^2}$ (B) $\dfrac{Gm^2}{4\pi\varepsilon_0 e^2}$ (C) $\dfrac{e^2}{4\pi\varepsilon_0 Gm^2}$ (D) $\dfrac{Gm^2 e^2}{4\pi\varepsilon_0}$ **[JAMB]**

Answer (C) $\dfrac{e^2}{4\pi\varepsilon_0 Gm^2}$

How?

16

The expression for gravitational force is given by $F_G = \dfrac{Gm^2}{r^2}$.

That of electrostatic force is given by $F_E = \dfrac{1}{4\pi\varepsilon_0} \dfrac{e^2}{r^2}$

$$\frac{F_E}{F_G} = \frac{e^2}{4\pi\varepsilon_0 r^2} \times \frac{r^2}{Gm^2} = \frac{e^2}{4\pi\varepsilon_0 Gm^2}$$

Therefore, option C is correct.

Question 5

A planet has mass m_1 and is at a distance r_1 from the sun. A second planet has mass $m_2 = 10m_1$ and is at a distance of $r_2 = 2r_1$ from the sun. Determine the ratio of the gravitational forces experienced by the planets.

(A) 1:5 (B) 2:5 (C) 3:5 (D) 4:5 **[JAMB]**

Solution!

The gravitational force of the sun on planet 1 is:

$$F_{sm1} = \frac{Gm_s m_1}{(r_1)^2} \quad \text{- - - - - -} (*)$$

The gravitational force of the sun on planet 2 is:

$$F_{sm2} = \frac{Gm_s m_1}{(r_2)^2} \quad \text{- - - - - -} (**)$$

But $m_2 = 10m_1$ and $r_2 = 2r_1$

Therefore, $F_{sm2} = \dfrac{Gms10m_1}{(2r_1)^2} \quad \text{- - - - -} (***)$

Dividing equation (*) by (***) gives

$$\frac{F_{sm1}}{F_{sm2}} = \frac{Gm_s m_1}{(r_1)^2} \times \frac{4r_1^2}{Gm_s 10m_1} = \frac{4}{10} = \frac{2}{5}$$

$\therefore$ The ratio is 2:5

| 19 |

Assuming the mass of the Earth to be M and the radius of the Earth R, the gravitational force of attraction of the Earth on object with mass (m) on the earth's surface is given by:

$$F = \frac{GMm}{R^2} \quad -------------------- 4$$

(Assume the earth is spherical in shape and has uniform density. So, the mass M is concentrated at the centre and the weight of the object, mg, = gravitational force).

$$mg = \frac{GMm}{R^2}, \qquad g = \frac{GM}{R^2} \quad ---------- 5$$

If g and G are accurately known, we can use equation (5) to calculate mass of the earth, M.

$$M = \frac{gR^2}{G} \quad ------------------- 6$$

Question 6

| 20 |

Two bodies have masses in the ratio 3:1. They experience forces which impact to them accelerations in the ratio 2:9 respectively. Find the ratio of the forces the masses experience.

(A) 1:4   (B) 2:1   (C) 2:3   (D) 2:5   **[JAMB]**

Solution!

| 21 |

$M_1 : M_2 \implies 3 : 1$

$a_1 : a_2 \implies 2 : 9$

$$\therefore \frac{F_1}{F_2} = \frac{M_1 a_1}{M_2 a_2} = \frac{3 \times 2}{1 \times 9} = \frac{6}{9} = \frac{2}{3}$$

$\therefore F_1 : F_2 = 2{:}3$

Question 7

| 22 |

What is the acceleration due to gravity g on the moon if g is 10ms^{-2} on the earth?

(A) 0.10ms^{-2} (B) 0.74ms^{-2} (C) 1.67ms^{-2} (D) 10.00ms^{-2} **[JAMB]**

Solution!

| 23 |

It is an established fact that the acceleration due to gravity on the moon is $\frac{1}{6} \times g$ that on the earth.

Therefore, on the moon g $= \frac{1}{6} \times \frac{10}{1} = 1.67\text{ms}^{-2}$.

5 Gravitational field strength

| 24 |

The gravitational field strength, $g^{|}$ at a point in a gravitational field is defined as the gravitational force of attraction per unit mass at that point.

i.e gravitational field strength, $g^{|} = \dfrac{gravitational\ force}{mass} = \dfrac{f}{m} = \dfrac{GM}{R^2}$ - - - - - 7

The gravitational field strength, $g^{|}$ is a vector and its S.I unit is Nkg^{-1}

Facts

| 25 |

From Newton's second law of motion, for a body of mass m in a gravitational field, the acceleration produced by the gravitational force is

Acceleration $= \dfrac{F}{M}$ which is equal to g.

The value of the gravitational field strength on the earth's surface is 9.8 NKg^{-1} and it produces an acceleration of 9.8 ms^{-2}.

Note: Acceleration due to gravity, g, and gravitational field strength, g', can be used interchangeably since they have the same S.I. unit.

6 Gravitational potential (V)

26

Another quantity associated with a gravitational field is the gravitational potential. The potential, v, at a point in a gravitational field is the work done in taking a unit mass from infinity to that point. It is a scalar quantity. The gravitational potential at infinity is assumed to be zero.

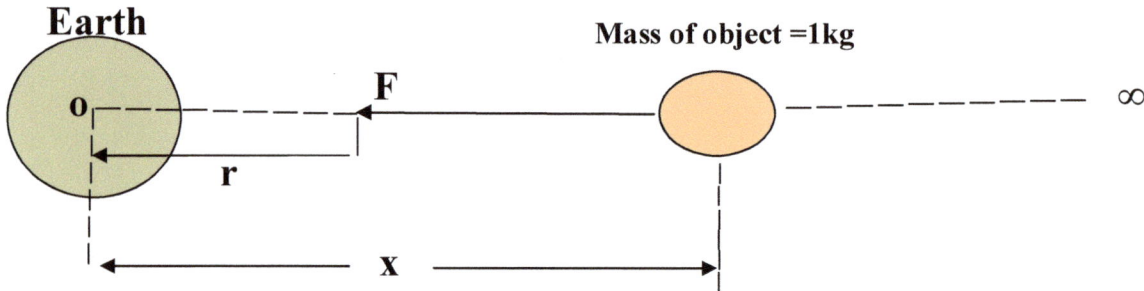

Figure 3. Gravitational potential on a unit mass, m

At a point r from the centre of the earth

$$V = -\frac{GM}{r} \quad \text{-------------------- 8}$$

The negative sign for V denotes that the gravitational potential at infinity is zero and decreases for points closer to the earth.

Facts

27

If a body is moved from infinity to the earth surface of radius R, the work done on the body is

$$V = -\frac{GM}{R} \quad \text{where R is the radius of the earth.}$$

From equation (8), it can be deduced that all points at the same distance from the centre of the earth have the same gravitational potential.

6.1 Equipotential surface

The surface where all points on it have the same gravitational potential is known as an equipotential surface.

The points lie on the surface of a sphere.

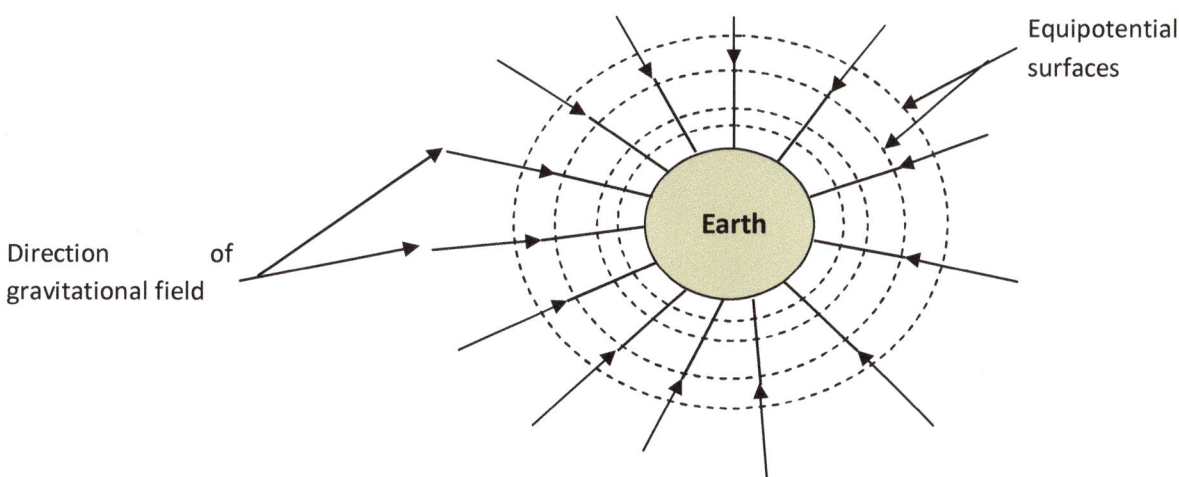

Figure 4. Equipotential surfaces

Fig 4 above shows equipotential surfaces around the earth. The direction of the field is normal to the equipotential surfaces.

6.2 Gravitational potential energy

The gravitational potential energy, U, of a body of mass m at a point in a gravitational field is defined as the work done to bring the body from infinity to that point.

i.e gravitational potential energy, $U = mV = -\dfrac{GMm}{r}$

$$V = -\frac{GM}{r} \text{ --- 9}$$

for a mass m at a distance r from earth which is of mass M.

7 Potential and kinetic energy of a satellite.

30

When a satellite of mass m, is moving round the earth of mass M, it has both kinetic energy (K.E) and potential energy (P.E). If it is moving round the earth with velocity, vms^{-1} and the radius of the orbit is r, then the force towards the centre of the orbit which is the centripetal force is equal to the gravitational force of attraction F.

That is, $\dfrac{mv^2}{r} = \dfrac{GMm}{r^2}$

$$\Rightarrow v^2 = \dfrac{GM}{r}$$

therefore, kinetic energy K.E is given by

$$\text{K.E} = \tfrac{1}{2}mv^2 = \dfrac{GMm}{2r} \text{ - 10}$$

7.1 Total (Mechanical) energy of the satellite

31

From plan 29, we have that P.E of mass in orbit = $-\dfrac{GMm}{r}$ (Assuming that the P.E in the earth's field at infinity is zero)

∴ Total energy in orbit = P.E + K.E.

$$= -\dfrac{GMm}{r} + \dfrac{GMm}{2r} = -\dfrac{GMm}{2r} \text{ - - - - - - - - - - - - - - - - 11}$$

7.2 Magnitude of energy of the satellite

32

From the equations (9, 10, and 11) above we can deduce that for the satellite:

(a) Magnitude of total energy (K.E +P.E) = magnitude of K.E

(b) K.E of the satellite increases as the radius of the orbit decreases.

(c) K.E of the satellite increases as the speed increases

(d) P.E of the satellite is numerically force of its K.E and of opposite sign.

8 Variation of g with latitude.

From the equation for the gravitational field strength,

$$g^{|} = \frac{F}{m} = g^{|} = \frac{GM}{r^2} \quad \text{-- ---------------------------------- 7}$$

on the surface of the earth (assume R= r).

We can see that if the earth is spherical, and G and M are constant, g will be constant over the earth's surface. But the earth is not a perfect sphere, so, R varies for different latitudes and g also varies. When R is small, g is large. In a nutshell, g varies with the latitude because earth is not spherical but ellipsoidal in shape and effect due to the rotational of the earth. Therefore, the value of g at the equator is lower than the value of g at the poles.

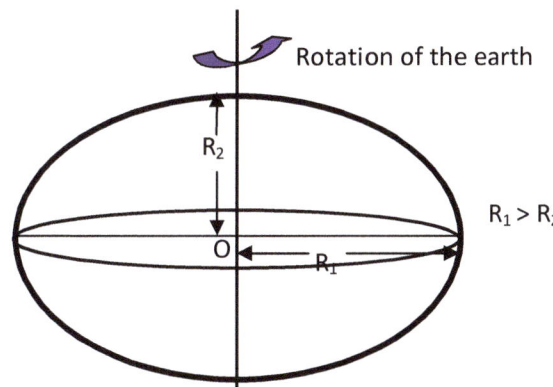

Figure 5. Variation of g with Latitude.

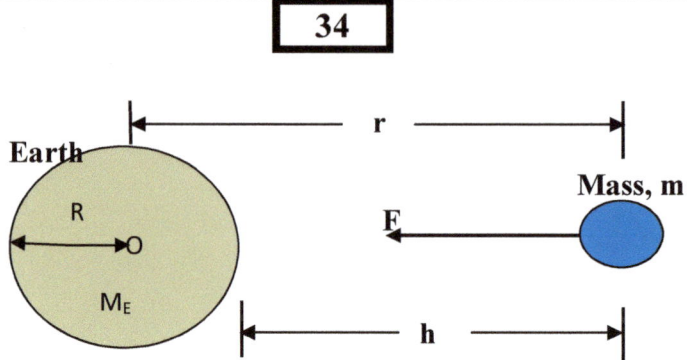

Figure 6. Variation of g with altitude (distance from the surface of the Earth)

The acceleration due to gravity g at a height h above the surface of the earth is given by $g^1 = \dfrac{GM}{r^2}$

Where $r = (R + h)$

And R = radius of the earth.

Therefore, $g^1 = \dfrac{GM}{(R+h)^2}$ - 12

Dividing equation (12) by (5), we get

$\dfrac{g^1}{g} = \dfrac{R^2}{(R+h)^2}$ - 13

Dividing the numerator and nominator right hand side of equation (13) by R^2, we get $\dfrac{g^1}{g} = \dfrac{1}{\left(1+\dfrac{h}{R}\right)^2}$.

If h is small compared to R, we can write

$g^1 = g\left(1+\dfrac{h}{R}\right)^{-2} = g\left(1-\dfrac{2h}{R}\right)$

$g^1 = g\left(1-\dfrac{2h}{R}\right)$ - 14

As seen from equation 14, the acceleration due to gravity g decreases with increasing altitude.

9.1 Deduction from equation (14)

35

From $g^1 = g(1 - \frac{2h}{R})$

When $h = \frac{1}{2}R$, $g^1 = 0$

This means that the gravitational pull of the earth on the body at that point is zero, and the body experiences a sensation of weightlessness.

Weightlessness is a feeling of having no weight and this happens when the gravitational pull of the earth has no effect on the body.

10 Escape velocity

36

We defined the gravitational potential energy of an object at a point to be the energy required to bring the object form infinity to that point.

Therefore, the gravitational potential energy of a mass m on the earth's surface is given by

$U = -\frac{GMm}{R}$, where M = mass of earth, R = radius of earth

The same amount of energy is required to take the same object from the surface of the earth to infinity.

10.1 Definition

37

Therefore, to launch a rocket of mass m from the earth's surface so that it escapes completely from the gravitational attraction of the earth, the rocket must take off from the earth with Kinetic energy $\geq \frac{GmM}{R}$

$\Rightarrow \frac{1}{2}mv^2 \geq \frac{GmM}{R}$

$\Rightarrow v \geq \sqrt{\frac{2GM}{R}}$

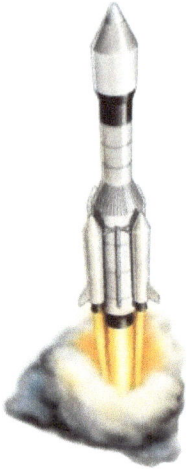

Figure 7. Rockets escaping Earth's gravity

The least velocity required is therefore $v = \sqrt{\dfrac{2GM}{R}}$, this velocity is called the escape velocity and with this velocity; the rocket will escape completely never to return.

Escape velocity is the least speed required for an object to escape completely from the earth and never to return.

10.2 Escape velocity from surface of the earth

The escape velocity from the surface of earth can, therefore, be calculated thus:

$$v = \sqrt{\frac{2GM}{R}}$$

but $GM = gR^2$ (from equation 5)

$$\therefore v = \sqrt{\frac{2gR^2}{R}} = \sqrt{2gR}, \quad g = 9.8\,\text{m/s}^2, R = 6.4 \times 10^6\,\text{m}$$

$$v = \sqrt{2 \times 9.81 \times 6.4 \times 10^6}$$

$$= 11.2 \text{ kms}^{-1}$$

Question 8

A body is projected from the earth's surface with the intension of letting it escape from the earth's gravitational field. What is the minimum escape velocity of the body? (given that earth's radius = 6.4 $\times 10^3 \, km$)

(A) 14kms^{-1} (B) 13kms^{-1} (C) 12kms^{-1} (D) 11kms^{-1} [JAMB]

See solution in plan 38 above.   **Answer: (D) 11kms^{-1}**

Question 9

If the radius of the earth is 6.4 $\times 10^6 \, m$, the escape velocity of a satellite from the earth is

(A) 1.13×10^4 m/s (B) 9.0×10^3 m/s (C) 8.0×10^3 m/s (D) 1.27×10^4 m/s

(g = 10m/s^2)   [JAMB]

Solution!

Solution:

g = 10m/s^2, R$_E$ = 6.4 $\times 10^6$ m

but $v = \sqrt{2gR_E}$ $= \sqrt{2 \times 10 \times 6.4 \times 10^6}$

v = 1.13 $\times 10^4$ m/s

11 Conservative forces

Observe that when you throw a ball into the air, as it moves up, the speed decreases. This is because its kinetic energy is being converted into potential energy.

Observe also that as it returns, the conversion is reversed and the speed of the ball increases as it falls down. This is because the potential energy is converted to kinetic energy.

Any force that offers this opportunity for a two-way conversion between kinetic and potential energies is called a conservative force.

11.1 Examples of conservative forces

43

Some of the examples of conservative forces include:

    i.   Gravitational force

    ii.   Spring force

    iii.   Electric force

11.2 Characteristics of conservative forces

44

The following are some of the examples of conservative forces.

   i. It is path independent: The work done by a conservative force such as gravity depends on the end point, not in the specific path traversed.

   ii. Net work done in a closed path is equal to zero i.e. if the initial point is equal to the final point then there is no net work done.

12 Non-Conservative Forces

45

For a non-conservative field, there is loss in kinetic energy which cannot be recovered by reversing the direction of action. It does not offer a two-way conversion between kinetic and potential energies.

12.1 Examples of Non-Conservative forces

46

Some of the examples of non-conservative forces include: air resistance or fluid resistance and kinetic friction.

12.2 Characteristics of non-conservative forces

The following are some of the characteristics of non-conservative forces:

i. It is path dependent: The path through which it traversed before it got to the final point always contributes to the final result.

ii. The net work in a loop is not equal to zero since mechanical energy is not conserved.

Exercises

1. Two spheres of masses 100kg and 90kg respectively have their centers separated by a distance of 10m. Calculate the magnitude of the force of attraction between them (G = 6.70×10^{-11} Nm^{2}kg^{-2}).

(A) 6.70×10^{-9} N   (B) 6.30×10^{-7} N   (C) 6.03×10^{-9} N   (D) 6.70×10^{-10} N

[WAEC/SSCE]

2. The magnitude of the gravitational force between two particles 0.10m a part is 10N. If the distance between them is increased to 0.20m, calculate the magnitude of the new force.

(A) 40.0 N   (B) 20.0 N   (C) 5.0 N   (D) 2.5N

[WAEC/SSCE]

3. If the distance between two suspended masses 10kg each is tripled. The gravitational force of attraction between them is reduced by:

(A) one half   (B) one third   (C) one quarter   (D) one ninth

[JAMB]   [Hint: F = $\dfrac{Gm_1m_2}{r^2}$]

4. A force of 200N acts between two objects at a certain distance apart. The value of the force when the distance is halved is:

(A) 400N   (B) 200N   (C) 100N   (D) 800N

5. Two spheres of masses 5.0kg and 10.0kg are 0.3m apart. Calculate the force of attraction between them (G = 6.67×10^{-11} Nm^{2}kg^{-2}).

(A) 3.50×10^{-10} N   (B) 3.7×10^{-8} N   (C) 3.57×10^{-2} N   (D) 4.00×10^{-2} N

[JAMB]

6. The force experienced by an object of mass 60kg in the moon's gravitational field is 100N. What is the intensity of the gravitational field?

(A) $0.60Nkg^{-1}$ (B) $1.67Nkg^{-1}$ (C) $6.12Nkg^{-1}$ (D) $9.81ms^{-2}$

[JAMB]

7. Calculate the density of the earth if it is a perfect sphere.

(Take $G = 6.67 \times 10^{-11} Nm^2kg^{-2}$, $R = 6.4 \times 10^6$ m, $g = 9.8$ m/s^2, Volume of a sphere $= \frac{4}{3}\pi R^3$).

[Hint: equation (6) and $density = \frac{mass}{volume}$]

8. The magnitude of the gravitational attraction between the earth and a particle is 40N. If the mass of the particle is 6kg, calculate the magnitude of the gravitational field intensity of the earth on the particle.

(A) $24.0m/s^2$ (B) $2.4m/s^2$ (C) $6.67m/s^2$ (D) $0.67m/s^2$

9. The mass and weight of a body on earth are 8kg and 80N respectively. Determine the mass and weight of the body respectively on a planet where the pull of gravity is $\frac{1}{8}$ that on the earth.

(A) 8kg, 8N   (B) 1kg, 10N   (C) 64kg, 10N   (D) 8kg, 10N

[WAEC/SSCE]

[Hint: mass of object is constant in every planet but weight, W = mg, varies]

10. What is the least velocity required for an object to escape completely from the surface of the earth?

Answers

1. C

2. D

3. D

4. A

5. B

6. B

7. 5.48×10^3 kgm^{-3}

8. C

9. D

10. 11.2 km/s